WORLD WAR 2 FACT F

Axis Cruisers

ANTHONY J. WATTS

JANE'S PUBLISHING COMPANY
LONDON · SYDNEY

First published in Great Britain in 1979 by
Jane's Publishing Company
Macdonald and Jane's Publishing Group Limited
Paulton House, 8 Shepherdess Walk, London N1 7LW

ISBN 0 354 01203 7

Printed in Great Britain by
Purnell & Sons Ltd
Paulton (Bristol) and London

Data

Classes and technical data have been arranged in chronological order so that the reader can follow the development of the cruiser within individual navies. The classes have been divided into the two main types of cruiser – Heavy and Light. Details of vessels cancelled under the Washington Treaty limitations have not been included.

The tabular data has been arranged in columns in the following order and relates to the particulars of given classes on the outbreak of war in September 1939, or as first completed (or as designed, if not completed) during the war. 1. Class name; 2. Displacement (standard tons); 3. Dimensions metric (length overall, beam and draught at full load); 4. Dimensions imperial (length overall, beam and draught at full load); 5. Type of machinery/number of shafts and propellers; 6. Designed horsepower/maximum designed speed (knots) (Figures rarely achieved during commission); 7. Radius of action (nautical miles)/Speed (knots); 8. Oil fuel (tons); 9. Main belt armour (maximum amidships/minimum ends) in millimetres (inches); 10. Deck armour in millimetres (inches); 11. Turret armour face/sides/roof in millimetres (inches); 12. Magazine armour in millimetres (inches); 13. Conning tower armour in millimetres (inches). The figures quoted for armour protection are the maximum over vital areas and do not cover thinning in any particular area. Details of bulkhead armour etc., are not quoted. 14. Guns and torpedoes – bore in millimetres/calibre (Arrangement of mountings where known); 15. Number of seaplanes or floatplanes carried/catapults fitted; 16. Complement.

Details are given of the known armament changes that were effected during the war and the date (in some cases approximate) of their first appearance.

The technical history is arranged as follows:
1. Name; 2. Builder (the legend for these is given below); 3. Month of first commissioning under country where listed; 4. Fate. Space precludes listing full details of losses. Vessels noted as 'lost' were lost in action with enemy forces. For further explanation of fates see table of abbreviations below.

List of Builders

1.	A G Weser, Bremen	Germany
2.	Ansaldo, Genoa	Italy
3.	Blohm & Voss, Hamburg	Germany
4.	Brest Navy Yard	France
5.	Cantieri Navali Riuniti, Ancona	Italy
6.	Cantieri Navali Riuniti, Palermo	Italy
7.	Cantieri Riuniti dell'Adriatico, Trieste	Italy
8.	Cantieri del Tirreno, Riva Trigoso	Italy
9.	Deutsche Werke, Kiel	Germany
10.	Germania Werft, Kiel	Germany
11.	Harima	Japan
12.	Kaiserliche Werft, Wilhelmshaven	Germany
13.	Kawasaki, Kobe	Japan
14.	Kiang Nan	Japan
15.	Kure Navy Yard	Japan
16.	Lorient Navy Yard	France
17.	Mitsubishi, Nagasaki	Japan
18.	Mitsubishi, Yokohama	Japan
19.	Naval Dockyard, Castellammare di Stabia	Italy
20.	Navalmeccanica, Castellammare di Stabia	Italy
21.	Odero-Terni-Orlando, Leghorn	Italy
22.	Odero-Terni-Orlando, La Spezia	Italy
23.	Orlando, Leghorn	Italy
24.	Sasebo Navy Yard	Japan
25.	Schichau, Danzig	Germany
26.	Stabilimento Tecnico Triestino, Trieste	Italy
27.	Uraga Dockyard	Japan
28.	Wilhelmshaven Naval Dockyard	Germany
29.	Yokosuka Navy Yard	Japan

Abbreviations

A/A	Anti-aircraft
AP	Armour Piercing
BU	Broken Up
Canc	Cancelled
CTL	Constructive Total Loss
D	Diesel
DCT	Director Control Tower
DP	Dual Purpose
E	Electric
Fr	France
ft	feet
Ger	Germany
Gr	Greece
GT	Geared Turbines
HA	High Angle
HE	High Explosive
hp	horsepower
in	inches
kg	kilogrammes
LA	Low Angle
MC	Marine Casualty
mg	machine gun
mm	millimetres
NA	Not Available
Pdr	Pounder
QF	Quick Firing
RNZN	Royal New Zealand Navy
Ret	Returned
Sbdg	Shipbuilding
Sc	Scuttled
shp	shaft horsepower
Stc	Stricken
Sz	Seized
T	Turbines
TE	Triple Expansion
Trf	Transferred
TT	Torpedo Tubes
UK	United Kingdom
USA	United States of America
USSR	Soviet Union
VTE	Vertical Triple Expansion (Reciprocating)

HEAVY CRUISERS
GERMANY

Admiral Hipper Class
Admiral Hipper, *Blücher*

It was some years before the Reichsmarine decided to embark on the construction of heavy cruisers. Being restricted by the Versailles Peace Treaty and Naval Treaties and concentrating on building up the strength of her army and air force Germany did not consider her naval strength of paramount importance, except in so far as commerce warfare was concerned. Consequently such efforts as were spared for naval development concentrated on submarine construction and developing the so-called pocket battleships of the *Deutschland* class (see Fact File on Battleships in this series).

Although work on designing heavy cruisers did not start until the 1930s, Germany profited from watching closely heavy cruiser developments in other navies, particularly noting their shortcomings.

As with all German naval construction the emphasis was on designs for commerce raiding although it is interesting to note that these vessels reverted to geared turbines for propulsion rather than the diesels of the *Deutschland* class. Although the diesels enabled vessels to gain a considerable range advantage, those fitted in the *Deutschland* had already given cause for concern over their maintainability and reliability (essential features in lone commerce raiding tactics) and it was decided to accept the range penalties imposed by turbine machinery in favour of its supposedly more reliable operation.

The design work on the new heavy cruiser was completed in the early 1930s and two cruisers were ordered in 1934.

Being designed for lone raiding operations particular attention was paid to armour protection, and the main belt was extended right up to the bows. However, other areas were less well protected and for a heavy cruiser it could be argued that the 1¼ inch main, upper deck and conning tower armour was insufficient (in contemporary American cruisers deck armour was about 2-inches, and conning tower armour up to 8-inches). Turret armour, on the other hand, was possibly excessive (6½-inches) when one considers that in her role as a lone raider she ought to avoid encounters with heavy enemy ships.

One of the notable features of the design was the strength and layout of the light A/A. This was further enhanced by the provision of two HA directors subsequently improved early in the war by the addition of another two HA directors. The light A/A on the *Admiral Hipper* was further strengthened during the war by the addition of 35 × 20mm.

During the early war refit the *Admiral Hipper* was given an 'Atlantic Bow' (as in other German warships) and a distinctive cap to her funnel.

Although fitted with turbine machinery and high pressure boilers of a novel type, the latter gave continual trouble and the whole machinery system required continual close attention by a well trained crew.

The *Blücher* was lost early in April 1940 during the invasion of Norway. She was torpedoed while sailing through the Norwegian fjord at Drobak and shore batteries contributed to her loss.

The *Admiral Hipper* survived until nearly the end of the war when she was eventually scuttled at Kiel, the hulk being scrapped post-war. During the war the *Admiral Hipper* was actively engaged in the Norwegian campaign (sinking the British destroyer *Glowworm*) and early in 1941 carried out a number of commerce raiding patrols.

Admiral Hipper in wartime camouflage. *Real Photographs*

Prinz Eugen Class
Lützow, *Prinz Eugen*, *Seydlitz*

The design for these heavy cruisers evolved from the previous *Admiral Hipper* class and although externally similar, they differed in a number of details. The *Prinz Eugen* was in fact completed with the 'Atlantic Bow' and distinctive funnel cap and was slightly larger than the *Admiral Hipper*.

Armour protection was on the same scale in both classes with armament initially on the same scale, although during the war *Prinz Eugen* had the 37 mm A/A removed and 15 single 40 mm Bofors mounts added. By January 1942 five quadruple 20 mm mountings had been added to the eight singles which were removed not long after.

By the summer of 1942 *Prinz Eugen* was carrying six quadruple and sixteen twin 20 mm mountings, but the twin mountings had been reduced to two in number by the autumn of 1944.

Although the A/A on the *Admiral Hipper* had been fairly well sited it had been found that the fire of the 37 mm on the upper deck amidships tended to be masked by the aircraft crane.

Consequently this was moved further forward in the *Prinz Eugen* to open up the arcs of fire. This meant that the catapult had to be moved from its original position over the after end of the hangar to the forward end of the hangar.

Only the *Prinz Eugen* of this class was completed, *Lützow* being sold partially completed to the Soviet Union (see Fact File on Allied Cruisers) while plans were drawn up to convert the *Seydlitz* into an aircraft carrier. Work on the conversion progressed very slowly and by the end of the war the ship had been completed to flight deck level. The hull was then scuttled at Könisberg, to be subsequently raised by the Soviets, renamed *Poltava*, but never completed, being scrapped about 1950.

The *Prinz Eugen* sailed on her first commission in May 1941 in company with the battleship *Bismarck* on her ill-fated voyage. Following the sinking of the *Hood*, the *Prinz Eugen* parted company from *Bismarck* and reached France. She later joined the *Scharnhorst* and *Gneisenau* in their celebrated dash up the English Channel. Towards the end of the war she carried out bombardment of the Soviet forces advancing towards Germany.

She was surrendered at the end of the war and allocated to the USA, being finally expended in the atom bomb tests at Bikini.

Prinz Eugen. *Real Photographs*

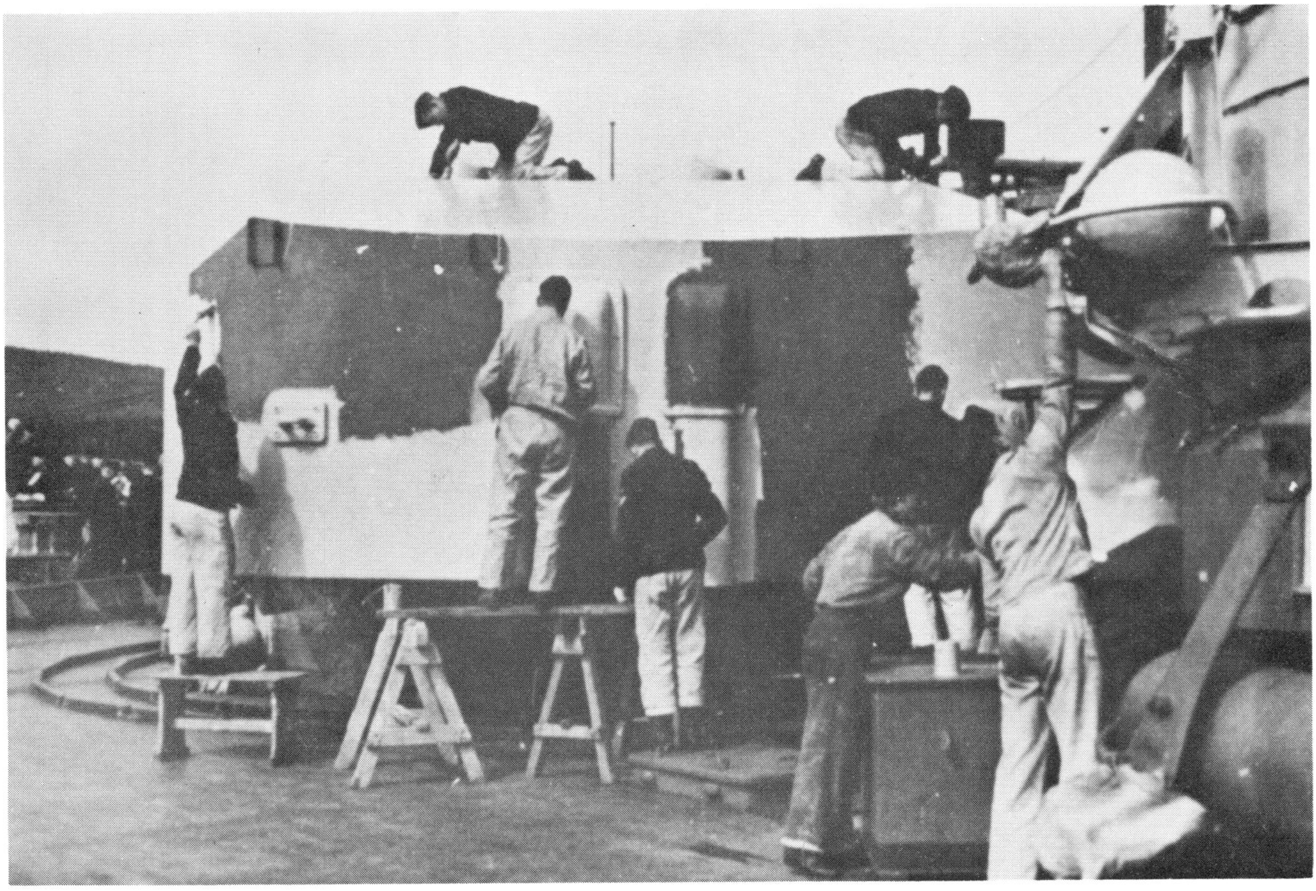

'A' turret on the *Prinz Eugen*. Preparations are being made for the fateful sortie with the *Bismarck*. *Author's collection*

Damage to the stern of *Prinz Eugen* received on February 23, 1942 when she was torpedoed by the British submarine *Trident*. *Real Photographs*

Prinz Eugen being escorted by *HMS Devonshire* after surrendering at Copenhagen at the end of the war. *IWM*

CLASS	*ADMIRAL HIPPER*	*PRINZ EUGEN*
Displacement	12,500	12,750
DIMENSIONS		
Metric	206 × 21.3 × 5.8	212.6 × 21.8 × 6.4
Imperial	675¾ × 70 × 19	697½ × 71¾ × 21
MACHINERY		
Unit/shafts	GT/3	GT/3
HP/speed	132,000/32½	132,000/32
Radius/speed	6,800/19	51,500/18
Fuel	4,250	4,250

CLASS	ADMIRAL HIPPER	PRINZ EUGEN
PROTECTION		
Main belt	70–82 (2¾–3¼)	70–82 (2¾–3¼)
Upper deck	12–32 (½–1¼)	12–32 (½–1¼)
Main deck	19–50 (¾–2)	19–50 (¾–2)
Turrets	70–108 (2¾–4¼)	70–108 (2¾–4¼)
Conning tower	19–152 (¾–6)	19–152 (¾–6)
ARMAMENT		
Guns & torpedo tubes	8 × 203 (4 × 2)	8 × 203 (4 × 2)
	12 × 104 (6 × 2)	12 × 104 (6 × 2)
	12 × 37 (6 × 2)	12 × 37 (6 × 2)
	4 × 20 (4 × 1)	8 × 20 (8 × 1)
	12 × 533 (3 × 4) TT	12 × 533 (3 × 4) TT
Aircraft/Catapults	3/1	3/1
Complement	1,600	1,600

NAME	BUILDER	IN SERVICE	FATE
Admiral Hipper	3	1939	Sc 3/5/45
Blücher	9	1939	Lost 9/4/40
Lützow	1	—	Trf USSR 1940, *Tallinn* qv
Prinz Eugen	10	1940	Expended as target 15/11/47
Seydlitz	1	—	Sc 10/4/45

ITALY

San Giorgio

The *San Giorgio* was an old armoured cruiser that had been completed before World War I. Although she was reconstructed by the naval dockyard at La Spezia between 1937–38 she was still considered to be of little fighting value and was deployed as a floating battery at Tobruk until scuttled there in January 1941 when the Italian forces abandoned the port.

The *San Giorgio* had a sister ship, the *San Marco*, which was converted to a radio controlled target ship in the early 1930s. This ship was taken over by the Germans in September 1943 when Italy surrendered and was found sunk in the harbour at La Spezia after the war.

Trento Class
Trento, Trieste

These two cruisers laid down in February 1925 (*Trento*) and June 1925 (*Trieste*) were, like other cruisers built at that time, subject to the limitations of the Washington Treaty. Like other navies, the Italians attempted to extract the maximum possible qualities out of a general design which basically was heavily restricted. The Italian solution to the problem was to sacrifice hull strength in an effort to provide some form of armour protection, inadequate though it was. The resultant design, although that of a heavy cruiser, was, nevertheless of extremely light construction with 3-inch side and turret armour.

The general design, like many other Italian designs of the period, although in essence satisfactory from the point of view that armour protection had been provided over vital areas and that a reasonable speed (in the region of 30 knots maximum) and a well disposed armament (eight 8-in (203 mm) in twin turrets and as first completed 16 × 3.9-in (100 mm) – four were subsequently removed) had been achieved, nevertheless suffered because of the lightness of construction. This point was well illustrated on numerous occasions during the war when, in inclement weather, the ships suffered structural damage when Allied units were unaffected. This disadvantage became manifest to such an extent that on a number of occasions the Italians either retired early from action, or avoided it altogether, in case, when manoeuvring at high speed, the hulls should suffer irreparable damage.

The cruisers carried three seaplanes which were launched from a catapult, sited rather strangely on the fo'c'sle deck.

Both cruisers were sunk during the war, the *Trento* by the submarine *Umbra* in the Ionian Sea in June 1942 and the *Trieste* by aircraft at Maddalena in April 1943.

Trieste in 1938. *M. Bar*

Trento. Note the 100 mm guns in twin beam mountings and the floatplane sited on the catapult on the foc's'le. *Author's collection*

A pre-war view of *Trento*. *Author's collection*

Zara Class
Fiume, Gorizia, Pola, Zara

These four heavy cruisers laid down between 1929–31 like the *Trento* class, sacrificed hull strength to gain adequate armour protection (it was increased from the 3-inches of the *Trento* class to $5\frac{1}{2}$-inches on the side armour and 5-inches on the turrets) and armament – this remained the same as in the *Trento*.

Although somewhat shorter than the *Trento* class displacement was increased, mainly as a result of the increased armour. Some weight saving was achieved by fitting lower powered machinery which reduced the maximum quoted speed by about 2 knots.

The cruisers were refitted in late 1937 when the 3.9-in (100 mm) guns abreast the after funnel were replaced by four sets of twin 37 mm. The 40 mm originally sited around the forefunnel were also removed about this time. A number of extra light A/A were added during the war.

Three of the class, *Fiume*, *Pola* and *Zara* were sunk by the British Mediterranean Fleet during the night action off Cape Matapan late in March 1941. *Gorizia* was under repair for bomb damage at La Spezia when Italy surrendered in September 1943. Attempts to scuttle the cruiser resulted in only superficial damage and the ship was seized by the Germans. Further repairs were not affected and the undocked ship was severely damaged by British and Italian frogmen and Chariots in June 1944. The hulk survived the war in a half sunken condition and was eventually scrapped.

Gorizia in 1933. *M. Bar*

Fiume in 1933. Note the size of the rangefinders and DCT, typical of Italian warship design. *M. Bar*

Zara in 1932. *Courtesy Aldo Fraccaroli*

Zara. Author's collection

Fiume. *Courtesy Aldo Fraccaroli*

The bridge of *Zara* in 1938. Note the aircraft on the catapult forward. *Courtesy Aldo Fraccaroli*

Zara leaving port. The floatplane on the foc's'le catapult is clearly visible. Note that the second and third 100 mm mounts are sited closer together than in the *Trento*. The deck has also been cutaway from aft of the bridge and flared inwards on the drop. *Author's collection*

Gorizia in 1938. *Courtesy Aldo Fraccaroli*

Gorizia in June 1946. Secondary and light A/A armament has been removed. *Courtesy Aldo Fraccaroli*

Bolzano following attack by frogmen. Note that the fore funnel has collapsed to port. *Real Photographs*

Bolzano

The *Bolzano* was built to virtually the same design as the *Trento* but with minor improvements. Dimensions, armament and machinery remained the same, but displacement rose by over 1,500 tons. Although the machinery was the same as in the *Trento* there was a considerable improvement in the maximum quoted speed which rose by 3 knots.

Like the *Trento* class, however, armour protection was still insufficient and hull strength was sacrificed in order to carry as powerful a machinery as possible in order to obtain a high speed.

The flush deck design of the *Trento* had not proved completely successful and so the raised fo'c'sle deck of the *Zara* class was adopted which resulted in the increase in displacement. Surprisingly in spite of extensive top hamper (notably the bridge and rangefinders) Italian cruisers, unlike the Japanese, did not suffer stability problems. As a result it was decided to reduce internal subdivision in the *Bolzano* and incorporate smaller magazines in an effort to keep displacement down to acceptable limits.

Like the other Italian heavy cruisers, she was refitted in 1937 when the aftermost twin 3.9-in (100 mm) were replaced by four twin 37 mm.

The cruiser was seized by the Germans at La Spezia when Italy surrendered and was finally sunk in that port by a combined Italian/British Chariot force launched from *MS 74* in June 1944.

CLASS	*SAN GIORGIO*	*TRENTO*	*ZARA*	*BOLZANO*
Displacement	9,500	10,500	11,500	11,065
DIMENSIONS				
Metric	140.9 × 21 × 6.9	197 × 20.6 × 6	182.8 × 20.6 × 6	197 × 20.6 × 6.1
Imperial	462¼ × 68¾ × 22½	646⅓ × 67½ × 19½	599¾ × 67½ × 19½	646¾ × 67½ × 20
MACHINERY				
Unit/shafts	TE/2	GT/4	GT/2	GT/4
HP/speed	18,000/23.2	150,000/35	95,000/32	150,000/36
Radius/speed				
Fuel				
PROTECTION				
Main belt	203 (8)	75 (3)	150 (6)	75 (3)
Deck	50 (2)	50 (2)	70 (2¾)	50 (2)
Turrets	203 (8)	100 (4)	150 (6)	100 (4)
Conning tower	248 (9¾)	100 (4)	150 (6)	100 (4)
ARMAMENT				
Guns & torpedo tubes	4 × 254/45 (2 × 2) 8 × 190/45 (8 × 1) 10 × 100/47 (10 × 1) 6 × 37 12 × 20 4 × 13.2	8 × 203/50 (4 × 2) 12 × 100/47 (6 × 2) 8 × 37 8 × 13.2 8 × 533 (2 × 4) TT	8 × 203/53 (4 × 2) 12 × 100/47 (6 × 2) 8 × 37 8 × 13.2	8 × 203/53 (4 × 2) 12 × 100/47 (6 × 2) 8 × 37 8 × 13.2 8 × 533 (2 × 4) TT
Aircraft/Catapults	—	3/1	2/1	3/1
Complement	NA	781	830	788

NAME	BUILDER	IN SERVICE	FATE	NAME	BUILDER	IN SERVICE	FATE
Bolzano	2	1933	Lost 21/6/44	San Giorgio	19	1910	Sc 22/1/41
Fiume	26	1931	Lost 28/3/41	Trento	23	1929	Lost 15/6/42
Gorizia	21	1931	Lost 26/6/44	Trieste	26	1928	Lost 10/4/43
Pola	21	1932	Lost 29/3/41	Zara	22	1931	Lost 29/3/41

JAPAN

Furutaka Class
Furutaka, Kako

Following the successful trials and evaluation of the novel design of *Yubari*, Vice Admiral Hiraga enlarged the concept of the small cruiser with heavy armament to a 7,000 ton light cruiser design carrying 7.9-in (200 mm) guns, a feature which no other light cruiser in the world could boast.

Much of the design of this class, prepared under the restrictions of the Washington Treaty, was based upon new lightweight materials and design techniques. Foremost among these features was the idea of utilising armour protection to form part of the hull strength members – the armour plating being built into the hull at an angle.

In addition to the radical change in the armour scheme the ships were fitted with very powerful machinery to give them a high speed. Like the *Naka*, these ships were completed with the new swan neck bow design which it was hoped would aid performance and improve seaworthiness. Coupled with this the fo'c'sle had a pronounced sheer and flare forward to assist in keeping the area around A and B turrets free from water.

They had a distinctive silhouette with a sleek racy look which was spoilt by the towering bridge structure, so common in many Japanese ships. Unfortunately the towering bridge structure, designed to house all the command and control facilities, gave rise to a certain instability owing to the massive topweight carried.

Owing to the weight restrictions occasioned by the Treaty the ships had to mount their heavy 7.9-in (200 mm) weapons in single light splinter shields rather than in multiple armoured turrets. These were all sited on the keel line three forward, with B gun superimposed over the other two, and two aft. The cruisers carried a heavy torpedo armament in fixed broadside firing positions aft of the boiler room. Abaft the after funnel was a flying off platform for the single seaplane that was carried; this was replaced soon after completion by a catapult.

The ships were extensively modernised between 1936–39, bulges being added to improve protection and give added stability together with a general re-distribution of weight. The major feature of the modernisation was that the cruisers were re-armed with six twin 8-in (203 mm) armoured turrets and the fixed torpedo tubes replaced by two quadruple trainable mounts abreast the catapult. The 3-in (76 mm) A/A were also removed and 4.7-in (120 mm) added. During this major reconstruction the ships were reboilered and the bridge rebuilt. All in all this increased the overall displacement to 9,000 tons.

Both vessels were active during the early stages of the war until lost in 1942, *Furutaka* being sunk by the American cruisers *Boise* and *Salt Lake City* off Savo Island in October and *Kako* torpedoed by the American submarine *S44* off Kavieng, during the battle for Savo Island in August.

Furutaka as completed in 1926. *Vicary*

Furutaka as modernised in 1939. Note that bridge has been extensively rebuilt, twin turrets have replaced single mounts and extra fire control positions added. *IWM*

Kako as completed in 1925. *Vicary*

Kako before modernization in 1930. *IWM*

Aoba Class
Aoba, Kinugasa

These two cruisers were improved versions of the *Furutaka* class ordered under the New Reinforcement Programme of 1923.

The main distinguishing feature between the two classes was that these two cruisers carried their 7.9-in (200 mm) guns in three twin turrets normally disposed, instead of the single shield mounts of the *Furutaka* class. In addition the catapult was moved further aft abaft the mainmast and 4.7-in (120 mm) A/A were carried instead of the 3-in (76 mm).

Like the *Furutaka's* these two cruisers were extensively modernised between 1938–40 when bulges were added and the ships re-gunned with 8-in (203 mm) weapons. The fixed torpedo tubes were also replaced by two quadruple, trainable mounts sited on the upper deck.

Aoba was severely damaged at the Battle of Cape Esperance, but managed to survive and returned home for repairs. The light A/A was increased to 15× 25 mm after the repairs and by June 1944 she carried 37× 25 mm. One of the torpedo tube mounts had been destroyed during the earlier action and was not replaced, the remaining mount being re-sited on the keel line. *Aoba* was further damaged while lying at Kure in July 1945 and the wreck was surrendered at the end of the war in a semi-sunken state.

Kinugasa was sunk off Rendova Island by carrier aircraft during the battle for Guadalcanal in November 1942.

Kinugasa as completed in 1926. *Vicary*

Aoba in 1935. *Vicary*

Aoba c.1935. *IWM*

Aoba c.1940. *IWM*

Ashigara Class
Ashigara, Haguro, Myoko, Nachi

This cruiser class was designed following experience gained with the *Yubari* and *Aoba* type designs. Lip service was still paid to the restrictions set down in the Washington Treaty, but when completed the new cruisers exceeded those limits.

Protection in this class was far superior to that in any other contemporary cruiser, the scheme of protection being a development of that installed in the *Aoba*. As in the *Aoba* inclined armour was used and its thickness increased by 1-inch over that in the *Aoba*. The total length of the main belt and deck armour was 410 feet and extended right over the machinery and boiler spaces. Increased protection against torpedo damage was also provided for in the new design and consisted of an arched protective bulkhead inside the main armour belt.

The four cruisers were laid down under the 1924 Programme and completed with five twin 7.9-in (200 mm) and six 4.7-in (120 mm) A/A. Like on the *Aoba* and *Furutaka* the torpedo tubes were in fixed broadside mountings, being sited in four groups of three abreast the catapult position.

Between 1934–36 the cruisers were given a major refit during which the 4.7-in guns, and fixed torpedo tubes were removed. Bulges were added both to improve stability and give added protection. These alterations enabled the upper deck to be extended from the fore funnel right aft to X-turret. On the new superstructure twin 5-in (127 mm) A/A together with new directors were added while trainable quadruple torpedo tubes were mounted aft of the catapults on the upper deck.

From 1938–1941 during a further extensive refit the turrets were regunned with 8-in (203 mm) barrels with much higher muzzle velocity. All control areas were modernised and two extra quadruple torpedo tube mounts added.

During May 1944 the light A/A was considerably increased to 24× 25 mm and by July 1944 a total of 52× 25 mm were carried and radar had been added. About this time the aircraft were removed.

Only the *Myoko* survived the war to be scuttled in the Malacca Strait. *Ashigara* was sunk in the Banka Straight by the submarine *Trenchant* in June 1945, *Haguro* by British destroyers off Penang in May 1945 and *Nachi* in an air attack off Corregidor in November 1944.

Myoko under construction in January 1929 with *Aoba* behind. Note the apertures for the fixed torpedo tubes in the side of the hull. *IWM*

Ashigara in 1929. *Vicary*

Ashigara in 1931. *IWM*

Ashigara in September 1931. *Author's collection*

Haguro. *Authors collection*

Ashigara in May 1937. Note that twin 127 mm mounts have replaced the single mounts originally fitted and that a large beam sponson has been built abreast the mainmast to carry the new A/A mount and the catapult. *Vicary*

Myoko in September 1931. *Author's collection*

Myoko on trials in March 1941 following further reconstruction. The bridge has been considerably enlarged, but the extensive plating round the sponson aft has been removed. *IWM*

Ashigara at the Spithead Coronation Review in May 1937. *Vicary*

The bridge of *Ashigara* in May 1937. *Authors collection*

Atago Class
Atago, Chokai, Maya, Takao

Laid down under the 1927 reinforcement programme these heavy cruisers were an improved *Ashigara* type. Visually the two classes were very similar, except that the *Atagos* had a much larger bridge structure. Although much larger than usual this bridge, was, nevertheless, fairly streamlined. Extensive use was made of electric welding to keep down weight and light alloys were used wherever possible.

Protection was increased in this class, particularly around the magazine. The ships were very heavily armed and the main armament had a maximum elevation of 70 degrees. By developing a main armament with an A/A capability it enabled the secondary A/A armament to be reduced to four guns resulting in sufficient weight saved to enable an extra 8-inch turret to be mounted.

The torpedo tubes were sited high up so that in the unfortunate event of an explosion there was little likelihood of watertight compartments being breached.

Atago and *Takao* were modernised in 1938 when additional superstructure aft was built on and the twin torpedo tubes replaced by quadruple mounts. A/A capability was further enhanced by replacing the 4.7-in (120 mm) weapons with twin 5-in (127 mm) DP guns. A new rangefinder and main gunnery director were also added. Beam was increased to 68 feet.

Chokai was never refitted in this manner while *Maya* only had a minor refit just before the war when 5-inch DP guns replaced the 4.7-inch weapons.

Following extensive air raid damage received in 1943 *Maya* was rebuilt, 'C' turret being replaced by two twin 5-inch A/A and 39× 25 mm added. The aircraft were also removed. Light A/A on *Maya* subsequently rose to 66× 25 mm before she was sunk by the submarine *Dace* in October 1944.

Atago and *Takao* carried a total of 26× 25 mm, *Atago* being sunk by the submarine *Dace* in October 1944 and *Takao* being severely damaged in Singapore harbour by frogmen from the midget submarine *XE-1* in July 1945. *Chokai* was severely damaged at the Battle of Leyte Gulf in October 1944 and had to be scuttled.

Takao in March 1932. *IWM*

Takao in July 1939. Note that new rangefinders and control positions have been added atop the bridge. *IWM*

Atago in August 1939. *IWM*

Mogami Class
Mikuma, Mogami, Kumano, Suzuya

These vessels were originally designed as light cruisers mounting five triple 6.1-in (155 mm) turrets. This arrangement was adopted in an effort to overcome the restrictions imposed by the Washington Naval Treaty, but it posed numerous problems for the design team.

To keep within the displacement limits severe restrictions on weights had to be imposed and great use was made of electric welding and light alloys.

The armament layout of three turrets forward with the third turret superimposed, and two aft was unmatched by any other cruiser and prompted the British and Americans to develop the *Southampton* and *Brooklyn* classes. In addition the *Mogami's* carried a heavy secondary A/A battery of eight 5-in (127 mm) DP guns. Twelve torpedo tubes in triple mounts were carried. Further weight saving was achieved by reducing the number of boilers while a new design of single geared impulse turbine was fitted.

Protection was on a similar scale to that in other heavy cruisers.

In spite of all the efforts put to saving weight etc. the ships showed signs of instability. Following gunnery trials with the first vessels signs of strain showed in certain of the welded joints. Work on the *Kumano* and *Suzuya* ceased; beam was increased and bulges added while these two cruisers were fitted with only eight boilers instead of the ten in *Mogami* and *Mikuma*.

In 1939 the class was completely rebuilt, the 6.1-in turrets being replaced by five twin 8-in (203 mm) and torpedo protection further improved by adding an extra, smaller, bulge.

Mikuma was sunk in air attacks during the Battle for Midway while the severely damaged *Mogami* returned home to have her stern rebuilt. The two after turrets were removed and a light flight deck supported by stanchions built out to the stern. It was planned that *Mogami* should then carry 11 aircraft. She recommissioned in May 1943 and carried a total of 30× 25 mm.

Kumano and *Suzuya* also received extra light A/A during 1943, carrying between 8–20× 25 mm. Early in 1944 this was further increased to 30× 25 mm. *Mogami* and *Suzuya* were abandoned and scuttled during the Battle for Leyte Gulf in October 1944 while *Kumano* was sunk in Manila during an air raid in November 1944.

Mogami on trials in 1935. Note that the rear 5-inch twin A/A and catapult has not been mounted. *IWM*

Mikuma in April 1939. *IWM*

Suzuya in 1939. *IWM*

Tone Class
Chikuma, Tone

These two cruisers were originally planned as improved *Mogami's* but considering the doubts expressed about the design of the *Mogami's* it was decided to omit one turret and mount only four triple 6.1-in (155 mm).

To save space all the main armament was mounted forward so as to leave a clear space aft for operating seaplanes as part of the role of these cruisers was to act as scouts for the Combined Fleet. Five seaplanes were to be carried and launched from catapults mounted on either quarter.

Extensive use of welding kept weight down but the problems encountered with the *Mogami's* convinced the designers that certain improvements necessitating an increase in displacement had to be incorporated. Bulges were added to improve stability and in the *Tone* these were built into the hull. Further armour plates were used as construction members to give added strength to the ships.

As in the *Atago* class the torpedoes were carried high up to minimise damage in case of accidental explosions.

The naval treaties expired before the ships had been gunned and it was decided to take advantage of this and substitute 8-in (203 mm) guns in place of the 6.1-in originally planned.

In 1943 A/A was increased to 20× 25 mm and after June 1944 to 56 × 25 mm.

Chikuma was severely damaged by aircraft at the Battle for Leyte Gulf in October 1944 and had to be scuttled while *Tone* finally sank from damage accumulated in air raids on Kure harbour towards the end of the war.

Ibuki

This was the last of the heavy cruisers to be designed for the Japanese Navy. It was based upon the modified *Suzuya* of the *Mogami* class. The design of the *Ibuki* embodied all the modifications that were subsequently built into the *Mogami* class.

Two vessels were ordered under the 1941 Programme and were planned with 8-in (203 mm) guns and quadruple 24-in torpedo tubes (as opposed to the triple mounts in the *Mogami* class). Machinery was to be identical to that fitted in the*Mogami* class.

After the *Ibuki* was launched in 1943 work on her ceased and it was decided in the summer of 1943 to convert her to an aircraft carrier. Work on the second vessel laid down in 1942 ceased altogether following the Battle of Midway. The assembled material was then dismantled to enable the carrier *Amagi* to be laid down on the vacated berth.

CLASS	*FURUTAKA*	*AOBA*	*ASHIGARA*	*ATAGO*
Displacement	8,700	9,000	13,380	13,400
DIMENSIONS				
Metric	185 × 16.5 × 5.5	185.2 × 15.8 × 5.7	203.7 × 20.7 × 6.3	203.7 × 18 × 6.5
Imperial	607 × 54 × 18¼	607½ × 51¾ × 18¼	668¼ × 68 × 20¾	668¼ × 59 × 21
MACHINERY				
Unit/shafts	GT/2	GT/4	GT/4	GT/4
HP/speed	103,390/33	108,456/33½	130,250/33¾	133,100/34¼
Radius/speed	8,000/14	8,000/14	8,500/14	8,500/14
Fuel	2,000	2,000	2,470	2,600
PROTECTION				
Main belt	76 (3)	76 (3)	102 (4)	76–102 (3–4)
Main deck	35 (1½)	35 (1½)	63–127 (2½–5)	80 (3½)
Turrets	25 (1)	80 (3½)	80 (3½)	80 (3½)
ARMAMENT				
Guns & torpedo tubes	6 × 203/50 (3 × 2) 4 × 120/45 (4 × 1) 8 × 25 4 × 13.2 8 × 610 (2 × 4) TT	6 × 203/50 (3 × 2) 4 × 120/45 (4 × 1) 8 × 25 4 × 13.2 8 × 610 (2 × 4) TT	10/203/50 (5 × 2) 8 × 127/40 (4 × 2) 8 × 25 16 × 610 (4 × 4) TT	10 × 203/50 (5 × 2) 8 × 127/40 (4 × 2) 8 × 25 16 × 610 (4 × 4) TT
Aircraft/Catapults	1/1	1/1	3/2	3/2
Complement	604	625	773	773

CLASS	*MOGAMI*	*TONE*	*IBUKI*
Displacement	11,200	11,215	12,200
DIMENSIONS			
Metric	201.5 × 19.2 × 5.8	201.6 × 18.4 × 6.1	200.5 × 19.2 × 6
Imperial	661 × 63 × 19⅓	661½ × 60½ × 20½	658 × 63 × 19¾
MACHINERY			
Unit/shafts	GT/4	GT/4	GT/4
HP/speed	152,000/34.9	152,000/35¼	152,000/33
Radius/speed	8,150/14	9,000/18	8,150/18
Fuel	2,163	2,950	2,163
PROTECTION			
Main belt	100 (4)	100 (4)	NA
Main deck	35 (1½)	64 (2½)	NA
Turrets	80 (3½)	76 (3)	NA
ARMAMENT			
Guns & torpedo tubes	10 × 203/50 (5 × 2) 8 × 127/40 (4 × 2) 8 × 25 4 × 13.2 12 × 610 (4 × 3) TT	8 × 203/50 (4 × 2) 8 × 127/40 (4 × 2) 12 × 25 12 × 610 (4 × 3) TT	10 × 203/50 (5 × 2) 8 × 127/40 (4 × 2) 8 × 25 4 × 13.2 16 × 610 (4 × 4) TT
Aircraft/Catapults	3/2	5/2	3/2
Complement	850	850	NA

NAME	BUILDER	IN SERVICE	FATE
Aoba	17	1927	Lost 28/7/45
Ashigara	13	1929	Lost 8/6/45
Atago	15	1932	Lost 23/10/44
Chikuma	17	1939	Lost 25/10/44
Chokai	17	1932	Lost 25/10/44
Furutaka	17	1926	Lost 11/10/42
Haguro	17	1929	Lost 16/5/45
Ibuki	15	—	Completed as carrier
Kako	13	1926	Lost 10/8/42
Kinugasa	13	1927	Lost 14/11/42
Kumano	13	1937	Lost 25/11/44
Maya	13	1932	Lost 23/10/44
Mikuma	17	1935	Lost 6/6/42
Mogami	15	1935	Lost 25/10/44
Myoko	29	1929	CTL 13/12/44
Nachi	15	1928	Lost 5/11/44
Suzuya	29	1937	Lost 25/10/44
Takao	29	1932	CTL 31/7/45
Tone	17	1938	Lost 24/7/45
301	17	—	Canc

LIGHT CRUISERS

GERMANY

Emden

Being barred from engaging in the manufacture of arms following the World War I, Germany was forced to rely on her previous war experience when she began to re-arm in the 1920s. The first light cruiser designed for the new German Navy, the *Emden*, was therefore based on the last light cruiser design Germany had developed during the World War I.

The original design showed a vessel carrying her main armament in twin turrets mounted on the keel line. This was not adhered to, and in the absence of suitable twin turrets having been designed the *Emden* carried single mountings on the keel line. Following World War I practise single mountings were also mounted on either beam to make up the broadside designed had she carried twin turrets.

It was realised on the outbreak of war in 1939 that the *Emden* was not sufficiently well armed and protected to engage in modern combat, the main deficiencies being considered to be a lack of adequate A/A armament and deck armour to withstand air attack. She was thus relegated to a training role in the Baltic interspersed with a number of minelaying operations. The *Emden* was bombed and then scuttled near Kiel in May 1945. The wreck was raised and scrapped in 1947.

Emden before the war. *Drüppel*

Emden displaying zig-zag camouflage in April 1940. *M. Bar*

A midships view of *Könisberg* showing the after funnel and catapult with an He 60 floatplane. *Author's collection*

Köln Class
Karlsruhe, Köln, Könisberg

This was the second class of light cruisers to be designed for the Reichsmarine and was designed for a commerce raiding role. To give them the extended range necessary for such a role and to enable high speeds to be developed for chasing fast merchant ships (or escaping from heavy enemy forces) they were given a dual machinery system of turbine and diesel units, the first such arrangement carried by a German warship. With the economy of the diesels they would obtain the extended range required while the turbines would develop the power necessary to obtain a high speed when in the chase.

Unfortunately the design suffered from a heavy concentration of weight forward realised by the position of A turret and the bridge. This concentration of weight in the forward part of the ship proved prejudicial to their sea-keeping qualities, the ships tending to bury their bows in heavy seas.

The two after main turrets were staggered either side of the keel line to widen the forward arcs of fire on either side.

The heavy secondary A/A armament was carried in single mounts aft when completed, but these were subsequently combined into twin mounts sited on the keel line. Later an extra twin mount was added, the two forward mountings being moved out to a beam position.

Karlsruhe was given an extensive refit before the war when the funnels were given distinctive caps, a heavy tripod mainmast added round the after funnel, extra control positions added round the foremast and bulges added.

Karlsruhe was torpedoed by the submarine *Truant* and then scuttled by the torpedo boat *Greif* off southern Norway in April 1940 while *Könisberg* was sunk in Bergen by Fleet Air Arm aircraft during the Norwegian campaign. She was salvaged during 1943 but capsized in September 1944 and finally abandoned.

Köln was sunk in shallow water by American aircraft in Wilhelmshaven late in March 1945.

Köln. Note the He 60 floatplane on catapult. *Drüppel*

Leipzig

The *Leipzig* was built to an improved *Könisberg* design with beam increased by three feet. The after main turrets were resited on the keel line while the exhaust was trunked into a single large funnel amidships. The aircraft catapult was resited in front of the funnel together with a large crane for handling the aircraft.

The machinery installation was improved with the addition of a third centre-line shaft coupled to a more powerful diesel arrangement. Protection was also improved with the main belt being extended fore and aft.

During 1941 six of the torpedo tubes were removed. Following her collision with the *Prinz Eugen* in October 1944 when she was badly holed just forward of the funnel, the forward boiler was removed together with the remaining torpedo tubes and the cruiser employed on training duties in the Baltic. Two quadruple and two single 20 mm were also added.

The cruiser survived the war and was scuttled loaded with gas ammunition off Lister in June 1946.

Nürnberg

The light cruiser *Nürnberg* was a further development of the *Köln* class with a slight increase in size and displacement over the *Leipzig*. Protection and armament remained almost identical but the main armoured belt was extended to the stern while the heavy A/A was centred amidships rather than being sited farther aft. The funnel was moved farther forward and the catapult resited aft of the funnel.

During the war the after two banks of torpedo tubes were

removed and A/A increased by the addition of two 37 mm and a total of 29× 20 mm, two quadruple, ten twin and a single mount.

After the capitulation she was allocated to the Soviet Navy and renamed *Admiral Makarov*.

Nürnberg. Note the catapult sited abaft the funnel. *Drüppel*

M-R

These light cruisers were designed especially to achieve high speed and improve the seaworthiness which had been found lacking in the earlier light cruisers. To attain these qualities topweight was reduced by carrying a lighter armament. The mixed diesel/turbine installation was retained to gain the advantage of extended range necessary for raiding operations.

Air defence, as with the earlier light cruisers, was, however, inadequate only a single HA director being mounted and only two twin heavy A/A mountings.

Protection was increased with the depth of the main belt being increased to encompass the machinery spaces up to the level of the upper deck and further extended between the outer barbettes.

They were designed for a dual role in mercantile warfare being equipped with a large mine magazine. With their high speed and adequate fire power they were well suited to the role for which the British Abdiel class had been designed.

Only *M*, *N* and *O* were laid down but little work had been done on the vessels before they were scrapped following Hitler's orders that no more large ships were to be built.

SP 1–3

When the previous class of cruisers was cancelled it left the major units without any form of screen protection, for the other light cruisers lacked both range and speed to operate with the heavy units.

Thus a new design – an enlarged destroyer – was prepared using the mixed diesel/turbine arrangement which had been proposed for the previous class, the diesels being coupled to the centre shaft. This increased the radius of action by over 30% above that of the *Bismarck* class battleships.

Armament in the new design was concentrated aft, indicating that they were not designed primarily to engage the enemy, but purely to act in a scouting role. Radar was to be fitted but quality was not up to that of the British and American sets.

Following the loss of the *Bismarck* and the timid role enacted by the *Tirpitz* the project was abandoned soon after the first unit had been laid down.

CLASS	*EMDEN*	*KÖLN*	*LEIPZIG*
Displacement	5,600	6,650	6,710
DIMENSIONS			
Metric	155 × 14.3 × 5.8	174 × 15.3 × 5.4	177 × 16.2 × 5.2
Imperial	508½ × 47 × 19	570¾ × 50¼ × 17¾	580¾ × 53¼ × 17
MACHINERY			
Unit/shafts	GT/2	GT+D/2	GT+D/3
HP/speed	46,500/29	68,000+ 1,800(D)/32	66,000+ 12,400(D)/32
Radius/speed	5,300/18	5,200/19	5,700/19
Fuel	1,266	1,406	1,531

CLASS	*EMDEN*	*KÖLN*	*LEIPZIG*
PROTECTION			
Main belt	50 (2)	50 (2)	50 (2)
Upper deck	19 (¾)	—	—
Main deck	19 (¾)	19 (¾)	19–25 (¾–1)
Turrets	13–19 (½–¾)	32 (1¼)	32 (1¼)
Conning tower	76 (3)	76 (3)	50 (2)
ARMAMENT			
Guns & torpedo tubes	8 × 150 (8 × 1) 3 × 90 (3 × 1) 4 × 37 4 × 533 (2 × 2) TT	9 × 150 (3 × 3) 6 × 90 (3 × 2) 8 × 37 (4 × 2) 12 × 533 (4 × 3) TT	9 × 150 (3 × 3) 6 × 90 (3 × 2) 8 × 37 (4 × 2) 10 × 20 12 × 533 (4 × 3) TT
Aircraft/Catapults	—	2/1	2/1
Complement	630	820	850

CLASS	*NÜRNBERG*	*M*	*Sp*
Displacement	6,980	7,800	4,542
DIMENSIONS			
Metric	181 × 16.3 × 5	183 × 16.3 × 6	152 × 14.6 × 4.5
Imperial	593¾ × 53¾ × 16½	600½ × 555¾ × 19¾	498¾ × 48 × 15
MACHINERY			
Unit/shafts	GT+D/3	GT+D/3	GT+D/3
HP/speed	60,000+12,400(D)/32	100,000+16,500(D)/35½	77,500+14,500(D)/36
Radius/speed	5,700/19	8,000/19	12,000/19
Fuel	1,448	1,600	1,352
PROTECTION			
Main belt	50 (2)	50 (2)	—
Upper deck	—	—	19 (¾)
Main deck	19–25 (¾–1)	25–38 (1–1½)	
Turrets	32 (1¼)	19–95 (¾–3¾)	25 (1)
Conning tower	76 (3)	50–100 (2–4)	—
ARMAMENT			
Guns & torpedo tubes	9 × 150 (3 × 3) 8 × 90 (4 × 2) 8 × 37 (4 × 2) 4 × 20 12 × 533 (4 × 3) TT	8 × 150 (4 × 2) 4 × 104 (2 × 2) 8 × 37 (4 × 2) 4 × 20 160 mines 8 × 533 (2 × 4) TT	6 × 150 (3 × 2) 2 × 90 (1 × 2) 8 × 37 (4 × 2) 12 × 20 140 mines 10 × 533 (2 × 5) TT
Aircraft/Catapults	2/1	2/1	—
Complement	896	920	583

NAME	BUILDER	IN SERVICE	FATE	NAME	BUILDER	IN SERVICE	FATE
Emden	28	1925	Sc 3/5/45	M	9	—	BU 1941–3
Karlsruhe	9	1929	Lost 10/4/40	N	28	—	BU 1941–3
Köln	28	1930	Lost 30/3/45	O	10	—	BU 1941–3
Könisberg	28	1929	CTL 22/8/44	P	10	—	Canc
Leipzig	28	1931	Sc 20/7/46	Q	25	—	Canc
Nürnberg	9	1935	Trf USSR 1946	R	9	—	Canc
			Admiral Makarov;	Sp 1	10	—	BU 1943
			Str 1961	Sp 2–3	10	—	Canc

ITALY

Alberico da Barbiano Class

Alberico da Barbiano, Alberto di Guissano, Bartolomeo Colleoni, Giovanni delle Bande Nere

Being designed for operations in the Mediterranean this class of light cruisers, all laid down in 1928, was designed for high speed and to carry the heaviest possible armament on a small displacement. To achieve these firing range and protection was sacrificed and like the *Trento* class heavy cruisers they were generally lightly built. These sacrifices at first did not seem of great importance in the Mediterranean where excessive range was not required and where it was considered that speed would enable the cruisers to keep out of trouble. Unfortunately the war proved this not to have been a good design feature for in action the ships suffered heavy battle damage and the light hull construction proved a considerable embarrassment in heavy weather.

These were the first light cruisers to be built for the Italian Navy after the World War I and are generally referred to as the *Condottieri* class.

Although they made their claimed maximum speed on trials this was achieved in the light condition and with some equipment not fitted. The main feature of the class, their high speed, was never in fact realised and by the outbreak of war their speed had dropped to such an extent that they were easily outclassed by the British cruisers.

All four cruisers were lost during the war, the *Alberico da Barbiano* and *Alberto di Guissano* being sunk off Cape Bon by the destroyers *Sikh*, *Maori*, *Legion* and *Isaac Sweers* in December 1941. The *Bartolomeo Colleoni* was sunk by the cruiser *Sydney* off Crete in July 1940 while the *Giovanni delle Bande Nere* was sunk by the submarine *Urge* off Stromboli in April 1942.

Alberico da Barbiano in 1931. *Author's collection*

Alberto di Guissano in 1939. *Courtesy Aldo Fraccaroli*

Luigi Cadorna Class
Armando Diaz, *Luigi Cadorna*

These two cruisers, laid down in 1930, were known as the Second Group of the *Condottieri* class, the first group being the previous class.

Dimensions remained the same as in the previous class but displacement was slightly increased, a result of an increase in the thickness of armour protection over certain areas. Machinery was the same as in the previous class, but the increase in displacement caused no appreciable loss in speed.

The main and secondary armament was the same as in the previous class, but the two vessels were provided with a better light A/A defence carrying a number of 20 mm. Like the previous class they were equipped for minelaying.

The *Armando Diaz* was sunk off Sfax by the submarine *Upright* in February 1941. *Luigi Cadorna* received minor changes to her armament during the war, the aircraft and catapult being removed in 1943, while the torpedo tubes were removed in 1944. She survived the war and served with the post-war Italian Navy until removed from the operational fleet in May 1951.

Luigi Cadorna c.1937. *Courtesy Aldo Fraccaroli*

Armando Diaz. Author's collection

Raimondo Montecuccoli Class
Muzio Attendolo, Raimondo Montecuccoli

These two cruisers were known as the Third Group of the *Condottieri* class. Although referred to as a *Condottieri* Type they were considerably larger than the previous two groups displacing 2,000 tons more than the Second Group and with a considerable increase in length. The increased size enabled more powerful machinery to be fitted, but the overall speed remained the same as in the previous types. Armour protection was considerably improved, the main belt being increased in thickness by nearly 2-in and other areas also being given thicker armour.

Like the previous group all the aircraft facilities and torpedo tubes were removed, this being effected on the *Raimondo Montecuccoli* in 1943. *Muzio Attendolo* had already been sunk in an air raid on Naples in December 1942. *Raimondo Montecuccoli* served in the post-war Italian Navy and was finally removed from active service in 1964.

Muzio Attendolo and *Raimondo Montecuccoli* in 1940. *Author's collection*

Muzio Attendolo c.1940–41. *Courtesy Aldo Fraccaroli*

Emanuele Filiberto Duca d'Aosta Class
Emanuele Filiberto Duca d'Aosta, Eugenio di Savoia

Known as the Fourth Group of the *Condottieri* Type these cruisers, whose basic design was the same as the previous types, again increased in size, length and beam being increased, much greater armour protection and consequently an increase in displacement. The increased size meant that yet again shaft horsepower had to be increased without achieving any speed advantage, there was in fact a slight drop of ½ knot in the designed speed.

Armament remained the same as in the previous types with the torpedo tubes being removed in 1943. Mine capacity was, however, increased the total number carried varied according to type.

The *Duca d'Aosta* survived the war and in February 1949 was transferred to the Soviet Union under the terms of the Peace Treaty. She was renamed *Stalingrad* (and subsequently *Kerch*) and finally disappeared from service about 1957–8 to be scrapped.

The *Eugenio di Savoia* also survived the war being removed from the Italian Navy list in June 1951 and transferred to the Greek Navy who renamed her *Helli*. She was discarded in 1964.

Luigi di Savoia Duca degli Abruzzi. Author's collection

Guiseppe Garibaldi Class
Guiseppe Garibaldi, Duca Degli Abruzzi

Known as the Fifth Group of the *Condottieri* Type designed as a counter to the French super-destroyers these two cruisers were not the same lightly armoured fast cruisers with a heavy armament that was characteristic of the previous groups.

The general silhouette remained the same – a long low vessel with widely spaced funnels, a distinguished circular tower bridge structure topped by an enormous fire control position and rangefinder.

Displacement and dimensions were increased yet again as was the overall scheme of armour protection.

Machinery was less powerful than the previous group and these ships sacrificed 2 knots in their maximum designed speed.

The main armament was increased to ten 6-in (152 mm), A and Y turrets being triple mounts with twin mounts superimposed in B and X positions. Torpedo armament was also increased by two tubes.

Laid down in 1933 both cruisers survived the war, the *Duca Degli Abruzzi* being discarded in 1961. *Guiseppe Garibaldi* survived another ten years and underwent a number of reconstructions before being deleted in 1971.

Luigi di Savoia Duca degli Abruzzi. *Author's collection*

Attilio Regolo Class

Attilio Regolo, Caio Mario, Claudio Druso, Claudio Tiberio, Cornelio Silla, Guilio Germanico, Ottaviano Augusto, Paolo Emilio, Pompeo Magno, Scipione Africano, Ulpio Traiano, Vipsanio Agrippa

Designed as high-speed light cruisers in reply to the French super-destroyers of the *Fantasque* class these cruisers were generally known as the *Capitain Romani* type.

The design was developed from the concept of the *Condottieri* type but with a maximum designed speed of 41 knots rather than the 37 knots of the *Condottieri* type. Work on the design began in 1937 and the majority of the class had been laid down by the end of 1939.

New armament was developed for the class, this being a 135 mm DP gun with a maximum range of 64,300 yards. The secondary armament comprised a 65 mm gun that was also specially developed for this class.

However, although it was planned that the cruisers should mount six of these guns they were never in fact fitted and the cruisers were armed with 37 mm instead.

Tonnage was severely limited in order to attain the desired speed and this resulted in virtually no armour protection being fitted, just some extra thin plating over the machinery spaces.

Although all the ships were laid down wartime shortages of essential materials resulted in only three being completed for service. *Claudio Druso*, *Claudio Tiberio*, *Paolo Emilio* and *Vipsanio Agrippa* were all suspended in June 1940 and the material assembled on the slips broken up between July 1941 and August 1942. Only the hull of the *Caio Mario* was delivered and this was scuttled at La Spezia in September 1943 to prevent her capture by the Germans. *Cornelio Silla* was 84% complete when work on her ceased. Her machinery was then removed and installed in the carrier *Aquila*. The hulk was seized by the Germans and sunk in an Allied air raid on Genoa in 1944. A similar fate befell the *Ottaviano Augusto*. The *Giulio Germanico* was seized by the Germans while under completion and they scuttled her just 17 days later. She was salvaged by the Italians and rebuilt between 1953–55 to serve in the new Italian Navy as the *San Marco*, being finally stricken in January 1971.

The *Ulpio Traiano* was launched in November 1942 and was under completion in Palermo when she was sunk by British Chariots in January 1943.

The *Attilio Regolo* and *Scipioni Africano* were both transferred to France after the war and renamed *Chateaurenault* and *Guichen*. They were rebuilt by the French and finally discarded in 1962 and 1961 respectively.

Pompeo Magno also survived the war and was rebuilt between 1953–55. She was renamed *San Giorgio* and is still in service.

Attilio Regolo in wartime camouflage. *Courtesy Aldo Fraccaroli*

The launch of *Scipione Africano* in January 1941. *Author's collection*

Scipione Africano in 1947. *Courtesy Aldo Fraccaroli*

Attilio Regolo after being torpedoed by the British submarine *Unruffled* on November 8, 1942. *Author's collection*

Etna Class
Etna (ex Taksin), Vesuvio (ex-Naresuan)

These two cruisers were laid down in 1939 for the Siamese Navy. They were taken over by the Italian Navy at the end of 1941 and their armament as designed altered to suit Italian requirements. Originally they were to have mounted six 6-in (152 mm), six 3-in (76 mm), eight 13.2 mm (or 20 mm), six 21-in torpedo tubes, two aircraft and a catapult. This was altered by the Italians to enable them to carry out the role of A/A cruisers escorting convoys to North Africa. In addition it was planned that they could themselves be used as fast transports on the North African run carrying troops, ammunition, supplies and fuel. Work proceeded very slowly, however, and when the Italians surrendered in September 1943 they were only 60% complete. The cruisers were seized by the Germans but never completed and the hulks were found sunk in Trieste harbour at the end of the war. They were refloated and scrapped in 1949.

Foreign cruisers
Taranto (ex-Strassburg)

This ex-German light cruiser was taken over by Italy in July 1920 and underwent a number of refits and modernisations. The major reconstruction took place between 1936–37 when she was adapted for Colonial service.

The cruiser was scuttled at La Spezia in September 1943 when Italy surrendered, but was salvaged by the Germans and sunk in an air raid in October 1943. The wreck was again salvaged but finally sunk in an air raid in September 1944.

Bari (ex-German *Pillau*, ex-Russian *Muraviev Amurski*)

This cruiser was originally ordered from Germany for the Russian Navy, but was taken over and completed by Germany on the outbreak of the World War I. She was handed over to Italy as war reparation in 1920 and rebuilt in 1934–5 for colonial service. It was planned to convert her to an A/A cruiser in the spring of 1943 but she was sunk in an air raid on Leghorn in June 1943 and the hulk further damaged by the Italians to prevent her use by Germany when Italy surrendered.

FR11, FR12

These two light cruisers were the ex-French *Jean de Vienne* and *La Galissonnière*. They had been scuttled at Toulon in November 1942 (see Fact File on Allied Cruisers) and were alloted to the Italian Navy by Germany. Italy refloated them but very little further work had been done on them by the time Italy surrendered in September 1943. Both vessels were retroceded in the same year.

Bari in 1925. *M. Bar.*

CLASS	*TARANTO*	*BARI*	*DA BARBIANO*
Displacement	3,184	3,248	5,200
DIMENSIONS			
Metric	138.7 × 13.5 × 4.7	135.3 × 13.6 × 4.9	169.3 × 15.5 × 4.9
Imperial	455 × 44⅓ × 15½	444 × 44½ × 16	555½ × 50¾ × 16
MACHINERY			
Unit/shafts	GT/2	GT/2	GT/2
HP/speed	25,000/27	28,000/27.5	95,000/37
Radius/speed			
Fuel			
PROTECTION			
Main belt	60 (2⅓)	85 (3⅓)	25 (1)
Deck	50 (2)	40 (1½)	20 (¾)
Turrets	50 (2)	50 (2)	25 (1)
Conning tower	100 (4)	75 (3)	40 (1½)
ARMAMENT			
Guns & torpedo tubes	7 × 150/43 (7 × 1) 2 × 76/40 (2 × 1) 8 × 20 10 × 13.2 120 mines	8 × 150/43 (8 × 1) 3 × 76/40 (3 × 1) 6 × 20 6 × 13.2 120 mines	8 × 152/53 (4 × 2) 6 × 100/47 (3 × 2) 8 × 37 8 × 13.2 4 × 533 (2 × 2) TT
Aircraft/Catapult	—	—	2/1
Complement	486	439	521

CLASS	*LUIGI CADORNA*	*MONTECUCCOLI*	*DUCA D'AOSTA*
Displacement	5,400	7,550	8,662
DIMENSIONS			
Metric	169.3 × 15.5 × 4.3	182.2 × 16.5 × 4.5	187 × 17.5 × 4.7
Imperial	555½ × 50¾ × 14	597¾ × 54 × 14¾	613½ × 57½ × 15½
MACHINERY			
Unit/shafts	GT/2	GT/2	GT/2
HP/speed	95,000/37	106,000/37	110,000/36.5
Radius/speed			
Fuel			
PROTECTION			
Main belt	42 (1½)	85 (3⅓)	105 (4¼)
Deck	20 (¾)	30 (1¼)	35 (1⅓)
Turrets	20 (¾)	70 (2¾)	100 (4)
Conning tower	50 (2)	100 (4)	100 (4)
ARMAMENT			
Guns & torpedo tubes	8 × 152/53 (4 × 2) 6 × 100/47 (3 × 2) 12–16 × 20 4 × 533 (2 × 2) TT	8 × 152/53 (4 × 2) 6 × 100/47 (3 × 2) 8 × 37 12 × 20 4 × 533 (2 × 2) TT 96 mines	8 × 152/53 (4 × 2) 6 × 100/47 (3 × 2) 8 × 37 12 × 20 6 × 533 (2 × 3) TT 100–146 mines
Aircraft/Catapults	2/1	2/1	2/1
Complement	544	650	694

CLASS	*GARIBALDI*	*REGOLO*	*ETNA*
Displacement	9,959	3,747	6,000
DIMENSIONS			
Metric	187 × 18.8 × 5.2	142.2 × 14.4 × 4	154 × 14.5 × 5.3
Imperial	613½ × 61¾ × 17	466½ × 47¼ × 13	505¼ × 47½ × 17⅓
MACHINERY			
Unit/shafts	GT/2	GT/2	GT/2
HP/speed	100,000/35	110,000/41	40,000/28
Radius/speed			
Fuel			
PROTECTION			
Main belt	130 (5)	?	60 (2⅓)
Deck	60 (2⅓)	?	30 (1¼)
Turrets	135 (5⅓)	?	—
Conning tower	140 (5½)	?	—
ARMAMENT			
Guns & torpedo tubes	10 × 152/55 (5 × 2) 8 × 100/47 (4 × 2) 8 × 37 10 × 20 6 × 533 (2 × 3) TT mines	8 × 135/45 (4 × 2) 8 × 37 8 × 20 8 × 533 (2 × 4)TT 70 mines	6 × 135/45 (3 × 2) 10 × 65 14 × 20
Aircraft/Catapults	2/2	1/–	—
Complement	892	420	570

NAME	BUILDER	IN SERVICE	FATE
Alberico da Barbiano	2	1931	Lost 13/12/41
Alberto di Giussano	2	1931	Lost 13/12/41
Armando Diaz	22	1933	Lost 25/2/41
Attilio Regolo	21	1942	Trf Fr 1948 *Chateaurenault* BU 1962
Bari	25	1914	BU 1943–4
Bartolomeo Colleoni	2	1932	Lost 19/7/40
Caio Mario	21	—	Sc 1943
Claudio Druso	8	—	BU 1942
Claudio Tiberio	21	—	BU 1941–2
Cornelio Silla	2	—	Lost 7/44
Emanuele Filiberto Duca d'Aosta	21	1935	Trf USSR 1949 *Kerch*; BU 1960s?
Etna	7	—	CTL 9/43
Eugenio di Savoia	2	1936	Trf Gr 1951, *Helli*, BU 1965
Giovanni delle Bande Nere	19	1931	Lost 1/4/42
Giuseppe Garibaldi	7	1937	In Service
Guilio Germanico	20	—	Sc 28/9/43
Luigi Cadorna	7	1933	BU 1951
Luigi di Savoia Duca Degli Abruzzi	22	1937	BU 1961
Muzio Attendolo	7	1935	Lost 4/12/42
Ottaviano Augusto	5	—	Lost 1/11/43
Paolo Emilio	2	—	BU 1941–2
Pompeo Magno	5	1943	*San Giorgio* 1955 BU 1971
Raimondo Montecuccoli	2	1935	BU 1964
Scipione Africano	21	1943	Trf Fr 1948, *Guichen*, BU 1961
Taranto	12	1912	Lost 23/9/44
Ulpio Traiano	6	—	Lost 3/1/43
Vesuvio	7	—	CTL 9/43
Vipsanio Agrippa	8	—	BU 1941–2
FR 11 salved ex-French		—	
FR 12 cruisers		—	

JAPAN

Tatsuta Class
Tatsuta, Tenryu

These two cruisers were designed following Japanese study of British World War I light cruisers, principally the *C* class. They were both ordered under the 1916 Programme and were the first modern second class light cruisers to be built for the Japanese Navy. The design of these ships laid the foundations for the design of subsequent classes of light cruiser for the Japanese Navy and remained constant until the middle 1920s when the concept of the light cruiser was omitted from future Japanese warship programmes.

Both vessels were laid down in 1917 and were the first Japanese warships to carry triple torpedo tubes. Completed in 1919 the cruisers were powered by British and American turbines, Parsons in the *Tenryu* and Curtis in the *Tatsuta*.

They achieved a good turn of speed at 33 knots, making them eminently suitable for use as flagships of destroyers, in which role they were employed until nearly the outbreak of war.

They underwent a major refit and modernisation in 1927 during which the single pole foremast was replaced by a tripod mast.

The ships were actively involved in the early Japanese successes at the start of the war and *Tenryu* was sunk by the American submarine *Albacore* off Madang in December 1942. Following this the *Tatsuta* returned to home waters, and being considered obsolete with outdated guns, fire control, and poor damage control was retained for local duties. She was sunk by the American submarine *Sandlance* off Yokosuka in March 1944.

Kitakami Class
Kiso, Kitakami, Kuma, Oi, Tama

These five cruisers laid down under the 8–4 Project of 1917 and ordered under the 1917–18 Programmes were similar in appearance and design to the previous class.

Displacement was increased by nearly 2,000 tons and the ships lengthened by 65 feet.

The increase in displacement and size would, it was hoped, improve seaworthiness. To retain the performance of the *Tatsuta* on the increased displacement it was necessary to fit more powerful machinery increasing horsepower from 51,000 to 90,000 which actually raised the speed by three knots. Bunkerage was also increased and the radius of action improved.

Protection remained much the same as in the *Tatsuta* while armament was improved by the addition of three 5.5-in (140 mm) mounts. The triple torpedo tube had not proved an unqualified success and so this class reverted to the old twin mounting.

The completion of *Oi* was delayed due to machinery failure experienced on full power trials and repairs delayed completion by nearly a year.

The ships were provided with aircraft facilities (a catapult and seaplane sited between two 5.5-in on the after deck) during the 1930s. *Kiso* differed from the other vessels in having an enlarged bridge structure with the seaplane hangar beneath the navigating bridge. She lacked the catapult.

During 1940 the ships were refitted with 24-in (533 mm) torpedo tubes as they became available while the 3-in (76 mm) A/A were replaced by 5-in (127 mm) DP weapons. Displacement rose to 5,610 tons and speed fell to 32¾ knots.

Oi and *Kitakami* were further re-armed in 1941–2 being fitted with ten quadruple 24-in torpedo tubes and up to 36× 25 mm in triple and single mounts. To carry the torpedo mounts the beam of the ships was sponsoned out and the alterations resulted in tonnage being increased to 5,870 tons. *Oi* was again re-armed in 1943, carrying an all DP armament of six 5-in and her light A/A increased to 36 × 25 mm.

Kitakami was also modified further, the eight after torpedo tube mounts being removed so that she could carry landing craft. Four of the torpedo tube mounts were replaced in 1944, however, while the four remaining 5.5-in were replaced by 5-in DP guns.

Following extensive torpedo damage inflicted by the submarine *Templar* on February 25, 1944, *Kitakami* had the after part from the aft machinery room rebuilt. All the torpedo tubes were removed and light A/A added, a total of 65 × 25 mm then being carried and radar added. The stern was constructed with a ramp and facilities for carrying and launching eight *Kaiten* suicide submarines provided. The after engine room was converted into a store room. This resulted in a drop in power from 90,000 to 30,000 shp and a reduction in speed to 23 knots.

Only *Kitakami* survived the war: *Kiso* was sunk by carrier aircraft off Manila in November 1944, *Kuma* was torpedoed by the submarine *Tally Ho* off Penang in January 1944, *Oi* by the American submarine *Flasher* in the Hong Kong area in July 1944 and *Tama* in the Philippine Sea area by the American submarine *Jallao* in October 1944.

Tama in 1923. *Vicary*

Oi before the war. *Author's collection*

Oi c.1930. *IWM*

Kuma as completed. *Vicary*

Kuma c.1935. Note catapult added and caps to funnels. *IWM*

Nagara Class
Abukuma, Isuzu, Kinu, Nagara, Natori, Yura

These light cruisers were an improvement on the previous class although structurally there was very little difference between the two classes. Displacement and dimensions remained the same but alterations were made to the layout of the interior.

A hangar for a seaplane was installed under the navigating bridge and in front of this projected a flying-off platform. Subsequently these aircraft facilities were removed and a catapult for the seaplane sited aft.

These cruisers were the first ships in the Japanese Navy to mount the 24-in (533 mm) torpedo tube which was carried in four twin tubes fore and aft on either beam.

Between 1934–5 the class underwent a major refit with tripod masts replacing pole masts and the removal of the seaplane to the after end of the vessels and the addition of the catapult.

During 1940 the 3-in (76 mm) A/A were replaced by two 5-in (127 mm) DP guns and in 1943 extra torpedo tubes added giving the ships a total of 24 tubes in six quadruple mounts. Light A/A was increased to 22 × 25 mm while two of the 5.5-in were removed. By 1944 the number of 25 mm had increased to 36 and the *Isuzu* was given extra A/A to enable her to operate as an A/A cruiser.

None of the class survived the war, three being sunk by American submarines and the other three by aircraft. *Isuzu* was torpedoed and sunk by *Charr* and *Gabilan* near the Soembawa Is. in April 1944, *Nagara* – by *Croaker* near Nagasaki in August 1944, and *Natori* – by *Hardhead*, also in August 1944. *Abukuma* was bombed and sunk by USAAF aircraft near Negros Is., *Kinu* – by USN aircraft near Marbata, while *Yura* was set afire by USN aircraft and then scuttled, all three cruisers being lost in October 1944.

Isuzu in 1921. *Vicary*

Nagara in August 1930. *Author's collection*

Nagara in September 1931. *Author's collection*

Kinu in September 1931. *Author's collection*

Kinu in September 1931. Note catapult extending in front of bridge. *Author's collection*

Kinu before refit. *IWM*

Kinu in January 1937. Note the catapult resited to just in front of 'X' turret. *IWM*

Sendai Class
Ayase, Jintsu, Minase, Naka, Otonase, Sendai

This was the last class of 5,000 ton light cruisers projected under the 8–8 Fleet Project. Again the basic hull design and dimensions remained the same as in earlier classes, but internal layout was modified. *Sendai* for instance was fitted with eleven oil-fired boilers and one mixed firing boiler compared to the eight light oil and four coal burners in *Nagara* and ten oil and two mixed firing boilers in the *Kuma*.

The improvement to the machinery installation in the *Sendai* led to one visible difference between the classes. This was occasioned by siting the mixed firing boiler in the *Sendai's* in the forward machinery space. This was exhausted through a tall fore funnel which was carried in addition to the other three funnels in the previous classes.

Naka, whose launching was delayed by earthquake was completed with the new swan neck bow with which all the other light cruisers were subsequently fitted.

The ships were completed with a seaplane hangar under the bridge and a flying off platform extending over A and B 5.5-in (140 mm). This was removed during the 1934–37 refit and modernisation and a catapult sited aft, one of the after 5.5-in being moved to the after end of the shelter deck to make the necessary space available.

The three vessels were further refitted during 1940 when the after twin torpedo tubes were removed and the forward ones replaced by quadruple mounts. Following the Battle of Midway in June 1942 the 3-in (76 mm) A/A were replaced by 5-in (127 mm) DP guns. Light A/A was also increased during the war, a total of 40 × 25 mm being carried, and one 5.5-in removed to compensate for the added topweight.

Jintsu and *Sendai* were both sunk by American surface forces, *Jintsu* at the Battle of Kula Gulf in July 1943 and *Sendai* at the Battle of Empress Augusta Bay in November 1943 both during the long drawn out campaign for the possession of Guadalcanal. *Naka* was sunk by carrier aircraft off Truk in February 1944.

Jintsu early in 1923. *IWM*

Naka in September 1931. *Author's collection*

Naka c.1934. Note the catapult moved aft but launching platform still retained. *IWM*

Jintsu in 1931. *Author's collection*

Jintsu in 1932. *Author's collection*

Yubari

The *Yubari* was originally planned under the 1917 Programme as the small cruiser *Ayase*. As such she was to have been used as a testbed for experimental machinery. The project failed to materialise and the order was transferred to the 1922 Estimates and the vessel named *Yubari*.

Even so the *Yubari* was planned as a trials ship to try out revolutionary concepts testing the limits to which cruiser designs could be stretched. Displacement was purposely restricted to 2,900 tons and within this parameter the Japanese attempted to design a vessel with the speed and firepower of a 5,000 ton cruiser.

If successful this novel concept would allow Japan to build nearly twice as many cruisers of a certain firepower as any other signatory of the Washington Treaty.

To achieve the heaviest possible broadsides *Yubari* carried her main armament of 5.5-in (140 mm) in two twin turrets superimposed over two single mounts in open shields fore and aft on the centreline. The aim proved successful, weight being saved by adopting two single open mounts at the expense of protection. The advantage of having twin turrets was that the crews were protected, unlike those in other light cruisers which only had open mounts. Three guns bearing astern was also considered to be an advantage, but this was countered in that only three guns covered the forward arcs.

Although the machinery was less powerful than in the 5,000 ton cruisers the 57,900 hp on three shafts still gave the vessel a speed equalling that of a 5,000 ton cruiser.

Yubari was refitted in 1927 and had larger shields attached to the torpedo tubes. During a further refit in 1943 the single 5.5-in and 3-in were removed and 4 × 25 mm A/A added and the 13 mm increased to eight.

She was torpedoed by the American submarine *Bluegill* off Palau in April 1944.

Yubari. Note the flush deck. *IWM*

Katori Class
Kashii, Kashima, Kashiwara, Katori, 815–818

These cruisers were designed in the early 1930s for training officer cadets in ship handling and control. Although small they were fitted with all the equipment to be found on larger vessels. A full gunnery control system was installed together with main and secondary A/A armament. A large catapult was also installed amidships.

The ships were completed with a large bridge incorporating a sizeable navigating area equipped with the latest navigational aids then available.

Both diesel and turbine machinery was fitted to enable engineer officers to become fully acquainted with the systems then operational in the Navy.

Kashima and *Katori*, both ordered under the 1930 programme, each carried out a single training cruise before the outbreak of war. They were later joined by the *Kashii* and entered the Fleet as fully operational flagships.

In 1941 the torpedo tubes were removed and two twin 5-in (127 mm) DP sited in the vacant space. Later 16× 25 mm were added and air and surface warning radar added. *Kashima* and *Kashii* were converted to A/S escorts in 1944, the aircraft and associated equipment being removed. On *Kashii* A/A was then increased to 25 × 25 mm and 300 depth charges added while *Kashima* carried 30 × 25 mm and only 100 depth charges.

Kashiwara ordered under the 1940 Programme was never completed owing to the imminent outbreak of war. Work ceased on her in November 1941 and the assembled material scrapped.

Kashii was sunk in an air raid off Cape Varela in January 1945 while *Katori* was sunk by American warships off Truk in February 1944. *Kashima* was surrendered.

Four vessels, Nos *815–818*, were planned under the 1942 Programme that was later cancelled. They were designed as escort cruisers armed with 8 × 3.9-in (100 mm) A/A guns.

Kashii in July 1941. *IWM*

Katori in April 1940. *IWM*

Agano Class
Agano, Noshiro, Sakawa, Yahagi

These light cruisers were designed in the late 1930s as a replacement for the obsolete 5,000 ton cruisers. A heavier armament of 6-in (152 mm) guns in triple turrets was to be mounted together with a much improved A/A capability of twin 25 mm. Six 24-in (533 mm) torpedo tubes in triple mounts were also to be fitted.

Armour protection was sacrificed to obtain the highest speed possible. The turrets, machinery and magazines were given just thin armour plating.

After studying the initial design the Navy altered the Staff requirement for the intended role and requested a high speed scout. The altered design replaced the 6-in with 5.9-in (150 mm) in three twin turrets. The weight saved by omitting X turret was used to mount two quadruple 24-in torpedo tubes with quick re-load gear sited amidships on the keel-line. Over the tubes a light platform was provided for the storage of two aircraft. A total of 16 torpedoes were provided for the tubes and in addition the vessels were fitted with sonar and carried 16 depth charges.

The light A/A was considerably augmented during the war, a total of 46 × 25 mm being carried in 1943, 52 × 25 mm in 1944, and by the autumn of 1944 the cruisers carried a total of 59 × 25 mm, as well as radar.

Noshiro and *Yahagi* were both sunk in air attacks, in October 1944 and April 1945 respectively, while *Agano* was torpedoed off Truk by the submarine *Skate* in February 1944. *Sakawa* was surrendered and expended in the Bikini atom test.

Noshiro in June 1943. *IWM*

Yahagi in December 1943. *IWM*

Oyodo Class
Oyodo, *Niyodo*

Known as the Type C this cruiser, although based on the design of the previous class, was planned for a completely different role. She was intended to act as a flagship of a combined squadron of scouting and attack type submarines and aircraft, the latter being reconnaissance planes of a new type specifically designed for operations from the *Oyodo*.

A total of six of these new high speed long-range aircraft were to be carried.

Oyodo herself was given extensive bunkerage for her role and was fitted with the latest communications and radar equipment. A large hangar aft was provided for the aircraft together with a very large catapult.

Protection was sacrificed in order to save weight and achieve the maximum possible speed. The main armament of triple turrets in A and B positions was obtained from the *Mogami* class when they were re-armed.

The *Oyodo* never operated in her intended role and the aircraft she was designed to carry were never built. Instead she carried two ordinary reconnaissance floatplanes. Late in 1943 *Oyodo* was refitted when a smaller catapult replaced the original one and the hangar converted to accommodation for the cruiser to serve as a flagship.

Light A/A was increased to 54 × 25 mm and radar added.

Oyodo saw little action during the war and was severely damaged in an air raid in March 1945. Temporary repairs were affected but the ship capsized in a further air raid on Kure harbour.

Type B
810–814, *5037–5038*

This design for a medium cruiser was prepared in 1941 and the ships were planned as high-speed cruisers to operate as squadron leaders for the *Shimakaze* class super destroyers. Five were ordered under the 1942 Programme that was later cancelled. This was replaced by the 1942 Modified Programme in which only two of these new cruisers were planned. These were also cancelled.

Foreign Vessels
Ihoshima Class
Ihoshima (ex-*Ning Hai*), *Yashojima* (ex-*Ping Hai*)

These two cruisers were built to a Japanese design for the Chinese Government. During the Sino-Japanese war both vessels were captured by the Japanese who placed them in the hands of the Nanking puppet Government who ran the country under Japanese direction.

Both cruisers were taken over by the Japanese Navy in 1943, and renamed. *Ihoshima* was sunk by the American submarine *Shad* off Honshu in November 1944 while *Yashojima* was sunk in an air raid on Luzon in the same month.

Ping Hai. Author's collection

CLASS	*TATSUTA*	*KITAKAMI*	*NAGARA*
Displacement	3,230	5,100	5,170
DIMENSIONS			
Metric	142.5 × 12.3 × 4	162.4 × 14.2 × 4.8	162.4 × 14.2 × 4.8
Imperial	468 × 40½ × 13	583 × 46¾ × 15¾	583 × 46¾ × 15¾
MACHINERY			
Unit/shafts	GT/3	GT/4	GT/4
HP/speed	51,000/33	90,000/36	90,000/36
Radius/speed	6,000/10	5,000/17	5,000/17
Fuel	900	1,610	1,450
PROTECTION			
Main belt	50 (2)	50 (2)	50 (2)
Deck	38 (1½)	37 (1½)	37 (1½)
Turrets	—	—	—
Conning tower	—	50 (2)	50 (2)

CLASS	*TATSUTA*	*KITAKAMI*	*NAGARA*
ARMAMENT			
Guns & torpedo tubes	4 × 140/50 (4 × 1) 1 × 80 6 × 533 (2 × 3) TT	7 × 140/50 (7 × 1) 2 × 80/40 (2 × 1) 8 × 533 (4 × 2) TT 80 mines	7 × 140/50 (7 × 1) 2 × 80/40 8 × 610 (4 × 2) TT
Aircraft/Catapults	—	—	1
Complement	332	439	438

CLASS	*SENDAI*	*YUBARI*	*KATORI*
Displacement	5,195	2,890	5,890
DIMENSIONS			
Metric	162.4 × 14.2 × 4.9	141 × 12 × 3.5	133.5 × 16 × 5.75
Imperial	583 × 46¾ × 16	463 × 39½ × 11¾	438 × 52⅓ × 18¾
MACHINERY			
Unit/shafts	GT/4	GT/3	GT + D/2
HP/speed	90,000/35¼	57,900/35½	8,000/18
Radius/speed	7,800/10	5,500/10	NA
Fuel	1,500	740	NA
PROTECTION			
Main belt	50 (2)	50 (2)	—
Deck	37 (1½)	25 (1)	50 (2)
Turrets	—	—	—
Conning tower	50 (2)	—	—
ARMAMENT			
Guns & torpedo tubes	7 × 140/50 (7 × 1) 2 × 80 8 × 610 (4 × 2) TT 80 mines	6 × 140/50 (2 × 2 + 2 × 1) 1 × 80/40 4 × 610 (2 × 2) TT 34 mines	4 × 140/50 (2 × 2) 2 × 127/40 (1 × 2) 4 × 25 4 × 533 (2 × 2) TT
Aircraft/Catapult	1	—	1/1
Complement	450	328	NA

CLASS	*AGANO*	*OYODO*	*TYPE B*	*IHOSHIMA*
Displacement	6,652	8,164	8,520	2,500 (normal)
DIMENSIONS				
Metric	174.5 × 15.2 × 5.5	192 × 16.6 × 5.95	NA	109.7 × 11.8 × 4.2
Imperial	572½ × 49¾ × 18½	630 × 54½ × 19½	NA	360 × 39 × 14
MACHINERY				
Unit/shafts	GT/4	GT/4	GT/4	GT/2
HP/speed	110,000/35	110,000/35	150,000/37½	9,000/22
Radius/speed	6,178/18½	10,619/18	6,178/18½	5,000/12
Fuel	1,405	2,360	NA	755
PROTECTION				
Main belt	57 (2¼)	50 (2)	NA	—
Deck	19 (¾)	50 (2)	NA	25 (1)
Turrets	25 (1)	63 (2½)	NA	25 (1)
Conning tower	—	—	—	—
ARMAMENT				
Guns & torpedo tubes	6 × 150/40 (3 × 2) 4 × 80/40 (2 × 2) 32 × 25 8 × 610 (2 × 4) TT	6 × 155/55 (2 × 3) 8 × 100/65 (4 × 2) 12 × 25 —	8 × 150/40 (4 × 2) NA NA NA	6 × 140/50 (3 × 2) 6 × 80/40 (3 × 2) 4 × 533 TT NA
Aircraft/Catapults	2/1	2/1	NA	2
Complement	730	NA	NA	340

NAME	BUILDER	IN SERVICE	FATE
Abukuma	27	1925	Lost 26/10/44
Agano	24	1942	Lost 26/2/44
Ayase	24	—	Canc 1922
Ihoshima	11	1943	Lost 19/11/44
Isuzu	27	1923	Lost 7/4/45
Jintsu	13	1925	Lost 13/7/43
Kashii	18	1941	Lost 12/1/45
Kashima	18	1940	BU 1947
Kashiwara	18	—	Canc BU 1942
Katori	18	1940	Lost 17/2/44
Kinu	13	1922	Lost 26/10/44
Kiso	17	1921	Lost 13/11/44
Kitakami	24	1921	BU 1947
Kuma	24	1920	Lost 11/1/44
Minase	27	—	Canc 1922
Nagara	24	1922	Lost 7/8/44
Naka	18	1925	Lost 17/2/44
Natori	17	1922	Lost 18/8/44
Niyodo	?	—	Canc
Noshiro	29	1943	Lost 26/10/44
Oi	13	1921	Lost 19/7/44
Otonase	24	—	Canc 1922

NAME	BUILDER	IN SERVICE	FATE	NAME	BUILDER	IN SERVICE	FATE
Oyodo	15	1943	Lost 28/7/45	Yahagi	24	1943	Lost 7/4/45
Sakawa	24	1944	Expended as target 2/7/46	Yashojima	14	1943	Lost 25/11/44
				Yubari	24	1923	Lost 27/4/44
Sendai	17	1924	Lost 2/11/43	Yura	24	1923	Lost 25/10/42
Tama	17	1921	Lost 25/10/44	810–814	—	—	Canc
Tatsuta	24	1919	Lost 13/3/44	815–818	—	—	Canc
Tenryu	29	1919	Lost 18/12/42	5037–5038	—	—	Canc

WORLD WAR 2 FACT FILES

This new series, prepared by experts who edit the world's foremost military, naval and aviation reference books, covers the fighting forces of the principal countries which took part in the Second World War. Each study provides a readable narrative supplemented by separate sections on technical details of weaponry that are seldom given in campaign and battle studies.

Published:

AIRCRAFT CARRIERS
AXIS SUBMARINES
ALLIED SUBMARINES
ALLIED CRUISERS
ANTI-TANK WEAPONS
MACHINE GUNS
BRITISH ESCORT SHIPS
AMERICAN GUNBOATS AND MINESWEEPERS
MORTARS AND ROCKETS
ANTI-AIRCRAFT GUNS
LIGHT AND MEDIUM FIELD ARTILLERY
SELF-PROPELLED ANTI-TANK AND ANTI-AIRCRAFT GUNS
HEAVY ARTILLERY
INFANTRY, MOUNTAIN AND AIRBORNE GUNS
AXIS PISTOLS, RIFLES AND GRENADES
ALLIED PISTOLS, RIFLES AND GRENADES
SUB-MACHINE GUNS AND AUTOMATIC RIFLES
AXIS COMBAT TANKS
ALLIED COMBAT TANKS

WW2 Aircraft Fact Files

U.S. NAVY AND MARINE CORPS FIGHTERS
JAPANESE ARMY AIR FORCE FIGHTERS (PARTS 1 AND 2)
SOVIET AIR FORCE FIGHTERS (PARTS 1 AND 2)
U.S. ARMY AIR FORCE FIGHTERS (PARTS 1 AND 2)
RAF FIGHTERS (PARTS 1 AND 2)
RAF BOMBERS (PARTS 1)

In preparation:

RAF FIGHTERS (PART 3)
RAF BOMBERS (PART 2)